THIS NIGHT SKY OBSERVATION REPORT BOOK

Belongs To:

Dedication

This Night Sky Observation Journal is dedicated to all the people out there who love watching the night sky and want to document their findings in the process.

You are my inspiration for producing books and I'm honored to be a part of keeping all of your observation notes and records organized.

This journal notebook will help you record your details about your night sky observations.

Thoughtfully put together with these sections to record: Date, Time, Location, Telescope, Conditions, Object, Finder, Field Sketch, & Notes.

How to Use this Book

The purpose of this book is to keep all of your Night Sky Observation notes all in one place. It will help keep you organized.

This Night Sky Observation Book will allow you to accurately document every detail about your night sky observation adventures. It's a great way to chart your course through watching the beautiful sky.

Here are examples of the prompts for you to fill in and write about your experience in this book:

1. Date, Time, Location

2. Telescope, Sky Conditions, Object

3. Finder

4. Field Sketch (Low Power, High Power, Eyepiece, Mag., Filter, FOV)

5. Notes (Blank Lined)

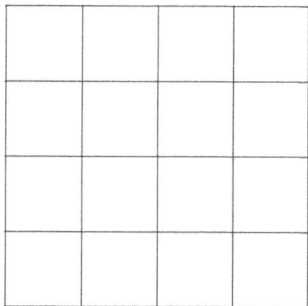

Finder

Date _____ Time _____

Location _____

Telescope _____

Sky Conditions _____

Object _____

Field Drawing

Low Power

High Power

Eyepiece:	Mag:
Filter:	FOV:

Eyepiece:	Mag:
Filter:	FOV:

Notes

_____

_____

_____

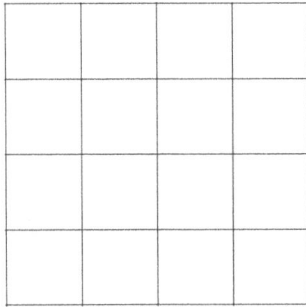

Finder

Date _____   Time _____

Location _____

Telescope _____

Sky Conditions _____

Object _____

Field Drawing

Low Power

High Power

Eyepiece:	Mag:
Filter:	FOV:

Eyepiece:	Mag:
Filter:	FOV:

Notes

_____

_____

_____

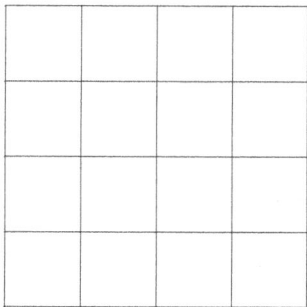

Finder

Date _____ Time _____

Location _____

Telescope _____

Sky Conditions _____

Object _____

Field Drawing

Low Power

High Power

Eyepiece:	Mag:
Filter:	FOV:

Eyepiece:	Mag:
Filter:	FOV:

Notes

_____

_____

_____

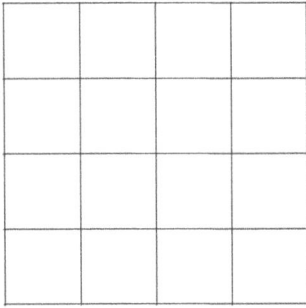

Finder

Date _____   Time _____

Location _____

Telescope _____

Sky Conditions _____

Object _____

Field Drawing

Low Power

High Power

Eyepiece:	Mag:
Filter:	FOV:

Eyepiece:	Mag:
Filter:	FOV:

Notes

_____

_____

_____

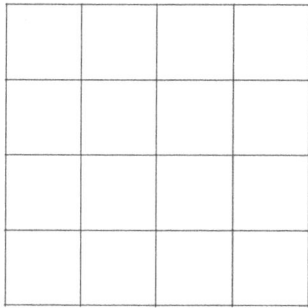

Finder

Date _____ Time _____

Location _____

Telescope _____

Sky Conditions _____

Object _____

Field Drawing

Low Power

High Power

Eyepiece:	Mag:
Filter:	FOV:

Eyepiece:	Mag:
Filter:	FOV:

Notes

_____

_____

_____

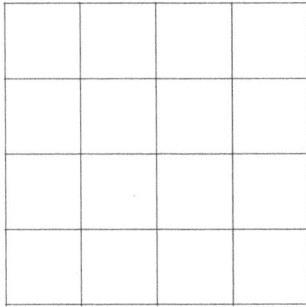

Finder

Date _____ Time _____

Location _____

Telescope _____

Sky Conditions _____

Object _____

Field Drawing

Low Power

High Power

Eyepiece:	Mag:
Filter:	FOV:

Eyepiece:	Mag:
Filter:	FOV:

Notes

_____

_____

_____

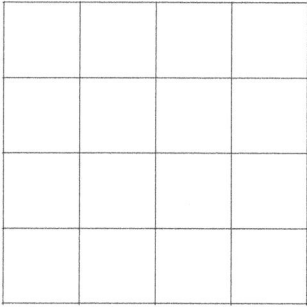

Finder

Date _____ Time _____

Location _____

Telescope _____

Sky Conditions _____

Object _____

Field Drawing

Low Power

High Power

Eyepiece:	Mag:
Filter:	FOV:

Eyepiece:	Mag:
Filter:	FOV:

Notes

_____

_____

_____

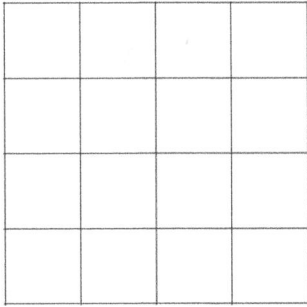

Finder

Date _____ Time _____

Location _____

Telescope _____

Sky Conditions _____

Object _____

Field Drawing

Low Power

High Power

Eyepiece:	Mag:
Filter:	FOV:

Eyepiece:	Mag:
Filter:	FOV:

Notes

_____

_____

_____

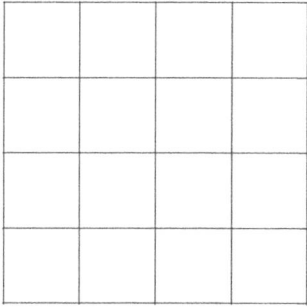

Finder

Date _____ Time _____

Location _____

Telescope _____

Sky Conditions _____

Object _____

Field Drawing

Low Power

High Power

Eyepiece:	Mag:
Filter:	FOV:

Eyepiece:	Mag:
Filter:	FOV:

Notes

_____

_____

_____

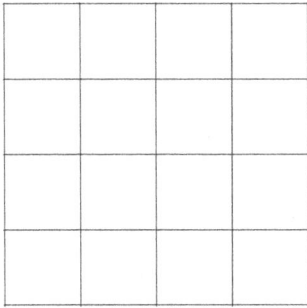

Finder

Date _____   Time _____

Location _____

Telescope _____

Sky Conditions _____

Object _____

Field Drawing

Low Power

Eyepiece:	Mag:
Filter:	FOV:

High Power

Eyepiece:	Mag:
Filter:	FOV:

Notes

_____

_____

_____

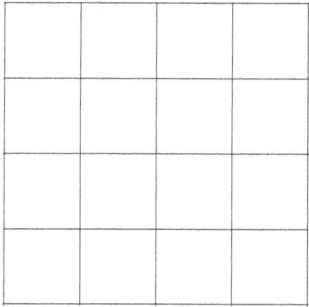

Finder

Date _____   Time _____

Location _____

Telescope _____

Sky Conditions _____

Object _____

Field Drawing

Low Power

High Power

Eyepiece:	Mag:
Filter:	FOV:

Eyepiece:	Mag:
Filter:	FOV:

Notes

_____

_____

_____

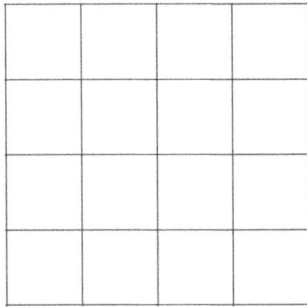

Finder

Date _____   Time _____

Location _____

Telescope _____

Sky Conditions _____

Object _____

Field Drawing

Low Power

High Power

Eyepiece:	Mag:
Filter:	FOV:

Eyepiece:	Mag:
Filter:	FOV:

Notes

_____

_____

_____

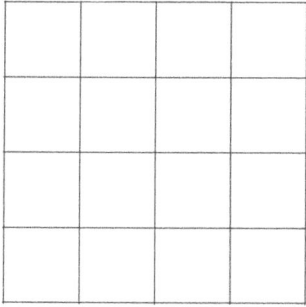

Finder

Date _____   Time _____

Location _____

Telescope _____

Sky Conditions _____

Object _____

Field Drawing

Low Power

High Power

Eyepiece:	Mag:
Filter:	FOV:

Eyepiece:	Mag:
Filter:	FOV:

Notes

_____

_____

_____

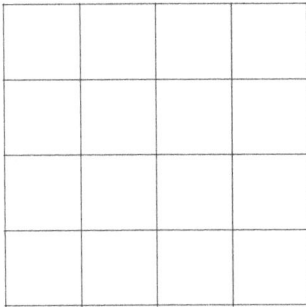

Finder

Date _____   Time _____

Location _____

Telescope _____

Sky Conditions _____

Object _____

Field Drawing

Low Power   High Power

Eyepiece:	Mag:
Filter:	FOV:

Eyepiece:	Mag:
Filter:	FOV:

Notes

_____

_____

_____

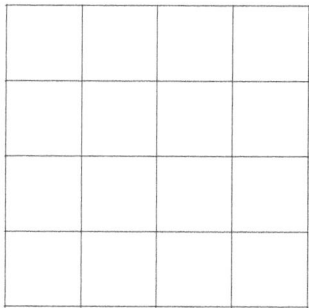

Finder

Date _____   Time _____

Location _____

Telescope _____

Sky Conditions _____

Object _____

Field Drawing

Low Power

High Power

Eyepiece:	Mag:
Filter:	FOV:

Eyepiece:	Mag:
Filter:	FOV:

Notes

_____

_____

_____

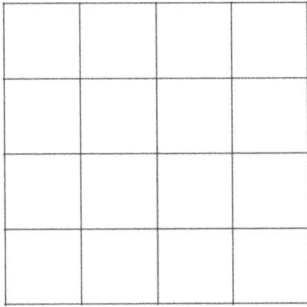

Finder

Date _____   Time _____

Location _____

Telescope _____

Sky Conditions _____

Object _____

Field Drawing

Low Power

Eyepiece:	Mag:
Filter:	FOV:

High Power

Eyepiece:	Mag:
Filter:	FOV:

Notes

_____

_____

_____

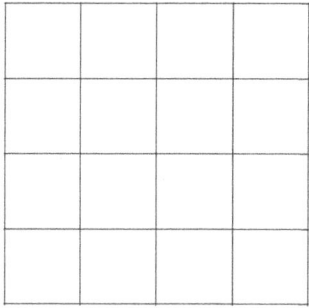

Finder

Date _____ Time _____

Location _____

Telescope _____

Sky Conditions _____

Object _____

Field Drawing

Low Power

Eyepiece:	Mag:
Filter:	FOV:

High Power

Eyepiece:	Mag:
Filter:	FOV:

Notes

_____

_____

_____

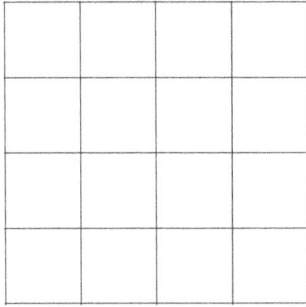

Finder

Date _____ Time _____

Location _____

Telescope _____

Sky Conditions _____

Object _____

Field Drawing

Low Power

High Power

Eyepiece:	Mag:
Filter:	FOV:

Eyepiece:	Mag:
Filter:	FOV:

Notes

_____

_____

_____

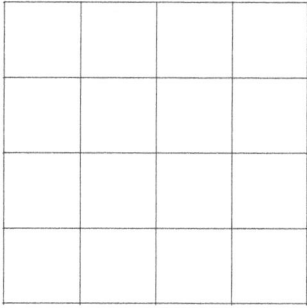

Finder

Date _____   Time _____

Location _____

Telescope _____

Sky Conditions _____

Object _____

Field Drawing

Low Power

High Power

Eyepiece:	Mag:
Filter:	FOV:

Eyepiece:	Mag:
Filter:	FOV:

Notes

_____

_____

_____

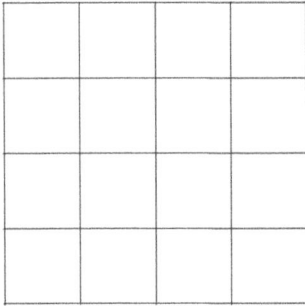

Date _____   Time _____

Location _____

Telescope _____

Sky Conditions _____

Object _____

Finder

Field Drawing

Low Power

High Power

Eyepiece:	Mag:
Filter:	FOV:

Eyepiece:	Mag:
Filter:	FOV:

Notes

_____

_____

_____

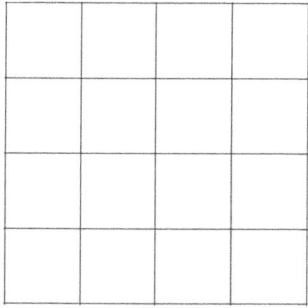

Finder

Date _____   Time _____

Location _____

Telescope _____

Sky Conditions _____

Object _____

Field Drawing

Low Power

Eyepiece:	Mag:
Filter:	FOV:

High Power

Eyepiece:	Mag:
Filter:	FOV:

Notes

_____

_____

_____

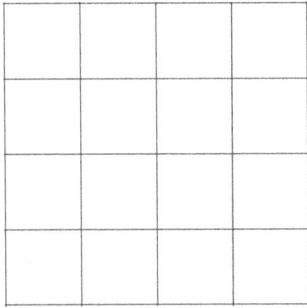

Finder

Date _____   Time _____

Location _____

Telescope _____

Sky Conditions _____

Object _____

Field Drawing

Low Power

High Power

Eyepiece:	Mag:
Filter:	FOV:

Eyepiece:	Mag:
Filter:	FOV:

Notes

_____

_____

_____

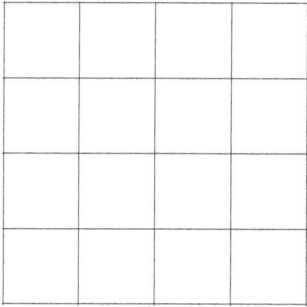

Finder

Date _____ Time _____

Location _____

Telescope _____

Sky Conditions _____

Object _____

Field Drawing

Low Power

High Power

Eyepiece:	Mag:
Filter:	FOV:

Eyepiece:	Mag:
Filter:	FOV:

Notes

_____

_____

_____

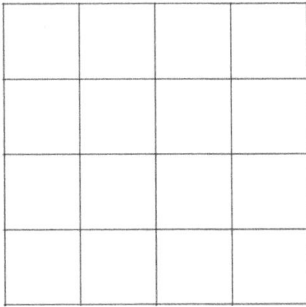

Finder

Date _____   Time _____

Location _____

Telescope _____

Sky Conditions _____

Object _____

Field Drawing

Low Power

High Power

Eyepiece:	Mag:
Filter:	FOV:

Eyepiece:	Mag:
Filter:	FOV:

Notes

_____

_____

_____

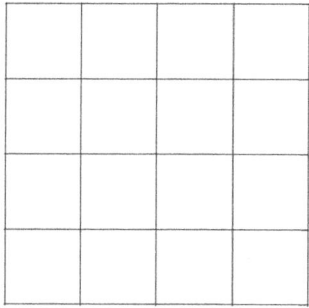

Finder

Date _____   Time _____

Location _____

Telescope _____

Sky Conditions _____

Object _____

Field Drawing

Low Power

High Power

Eyepiece:	Mag:
Filter:	FOV:

Eyepiece:	Mag:
Filter:	FOV:

Notes

_____

_____

_____

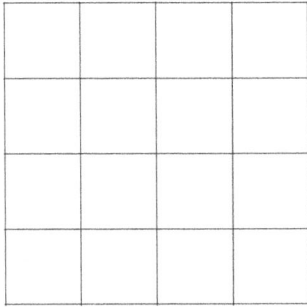

Finder

Date _____   Time _____

Location _____

Telescope _____

Sky Conditions _____

Object _____

Field Drawing

Low Power

High Power

Eyepiece:	Mag:
Filter:	FOV:

Eyepiece:	Mag:
Filter:	FOV:

Notes

_____

_____

_____

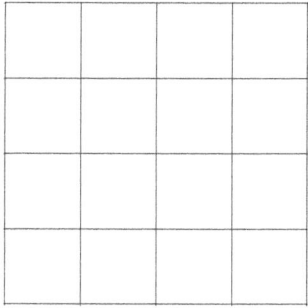

Finder

Date _____ Time _____

Location _____

Telescope _____

Sky Conditions _____

Object _____

Field Drawing

Low Power

High Power

Eyepiece:	Mag:
Filter:	FOV:

Eyepiece:	Mag:
Filter:	FOV:

Notes

_____

_____

_____

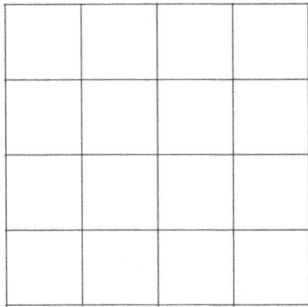

Finder

Date _____ Time _____

Location _____

Telescope _____

Sky Conditions _____

Object _____

Field Drawing

Low Power

High Power

Eyepiece:	Mag:
Filter:	FOV:

Eyepiece:	Mag:
Filter:	FOV:

Notes

_____

_____

_____

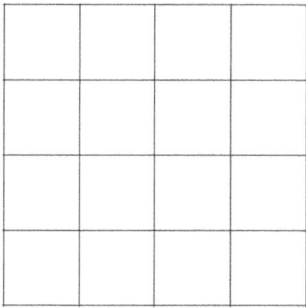

Finder

Date _____   Time _____

Location _____

Telescope _____

Sky Conditions _____

Object _____

Field Drawing

Low Power

Eyepiece:	Mag:
Filter:	FOV:

High Power

Eyepiece:	Mag:
Filter:	FOV:

Notes

_____

_____

_____

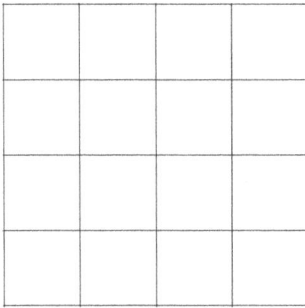

Finder

Date _____   Time _____

Location _____

Telescope _____

Sky Conditions _____

Object _____

Field Drawing

Low Power

High Power

Eyepiece:	Mag:
Filter:	FOV:

Eyepiece:	Mag:
Filter:	FOV:

Notes

_____

_____

_____

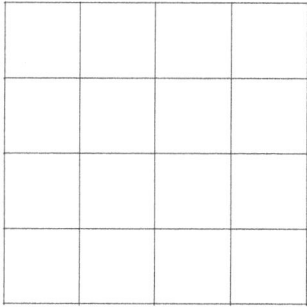

Finder

Date _____   Time _____

Location _____

Telescope _____

Sky Conditions _____

Object _____

Field Drawing

Low Power

High Power

Eyepiece:	Mag:
Filter:	FOV:

Eyepiece:	Mag:
Filter:	FOV:

Notes

_____

_____

_____

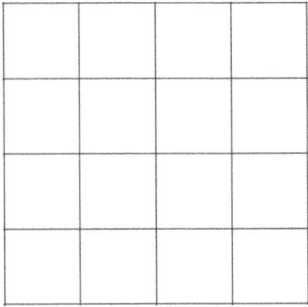

Finder

Date _____   Time _____

Location _____

Telescope _____

Sky Conditions _____

Object _____

Field Drawing

Low Power

High Power

Eyepiece:	Mag:
Filter:	FOV:

Eyepiece:	Mag:
Filter:	FOV:

Notes

_____

_____

_____

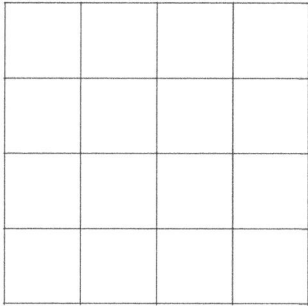

Finder

Date _____   Time _____

Location _____

Telescope _____

Sky Conditions _____

Object _____

Field Drawing

Low Power

High Power

Eyepiece:	Mag:
Filter:	FOV:

Eyepiece:	Mag:
Filter:	FOV:

Notes

_____

_____

_____

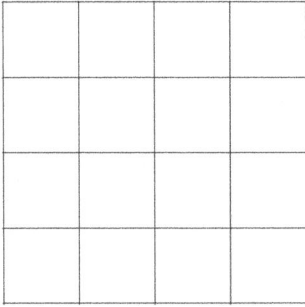

Finder

Date _____   Time _____

Location _____

Telescope _____

Sky Conditions _____

Object _____

Field Drawing

Low Power

High Power

Eyepiece:	Mag:
Filter:	FOV:

Eyepiece:	Mag:
Filter:	FOV:

Notes

_____

_____

_____

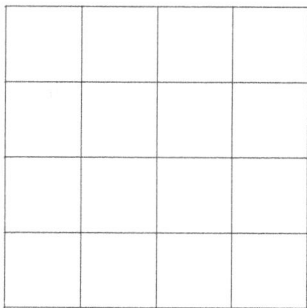

Finder

Date _____ Time _____

Location _____

Telescope _____

Sky Conditions _____

Object _____

Field Drawing

Low Power

High Power

Eyepiece:	Mag:
Filter:	FOV:

Eyepiece:	Mag:
Filter:	FOV:

Notes

_____

_____

_____

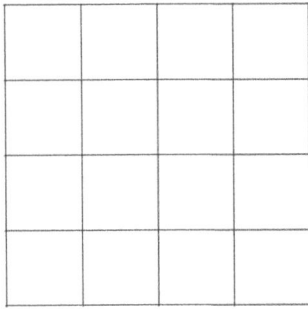

Finder

Date _____   Time _____

Location _____

Telescope _____

Sky Conditions _____

Object _____

Field Drawing

Low Power

High Power

Eyepiece:	Mag:
Filter:	FOV:

Eyepiece:	Mag:
Filter:	FOV:

Notes

_____

_____

_____

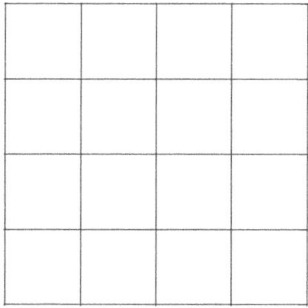

Finder

Date _____   Time _____

Location _____

Telescope _____

Sky Conditions _____

Object _____

Field Drawing

Low Power

High Power

Eyepiece:	Mag:
Filter:	FOV:

Eyepiece:	Mag:
Filter:	FOV:

Notes

_____

_____

_____

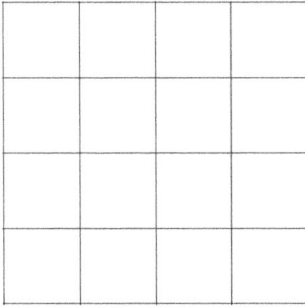

Finder

Date _____   Time _____

Location _____

Telescope _____

Sky Conditions _____

Object _____

Field Drawing

Low Power

High Power

Eyepiece:	Mag:
Filter:	FOV:

Eyepiece:	Mag:
Filter:	FOV:

Notes

_____

_____

_____

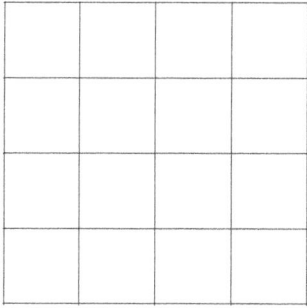

Finder

Date _____   Time _____

Location _____

Telescope _____

Sky Conditions _____

Object _____

Field Drawing

Low Power

High Power

Eyepiece:	Mag:
Filter:	FOV:

Eyepiece:	Mag:
Filter:	FOV:

Notes

_____

_____

_____

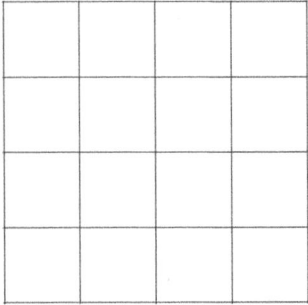

Finder

Date _____   Time _____

Location _____

Telescope _____

Sky Conditions _____

Object _____

Field Drawing

Low Power

High Power

Eyepiece:	Mag:
Filter:	FOV:

Eyepiece:	Mag:
Filter:	FOV:

Notes

_____

_____

_____

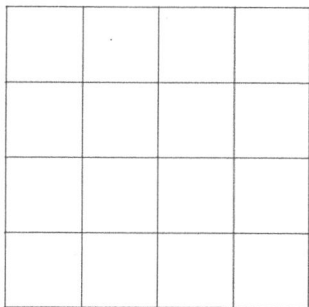

Finder

Date _____   Time _____

Location _____

Telescope _____

Sky Conditions _____

Object _____

Field Drawing

Low Power

High Power

Eyepiece:	Mag:
Filter:	FOV:

Eyepiece:	Mag:
Filter:	FOV:

Notes

_____

_____

_____

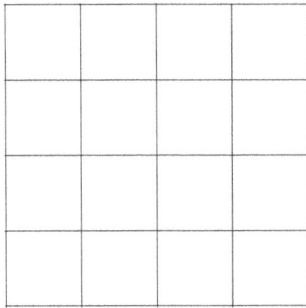

Finder

Date _____ Time _____

Location _____

Telescope _____

Sky Conditions _____

Object _____

Field Drawing

Low Power

High Power

Eyepiece:	Mag:
Filter:	FOV:

Eyepiece:	Mag:
Filter:	FOV:

Notes

_____

_____

_____

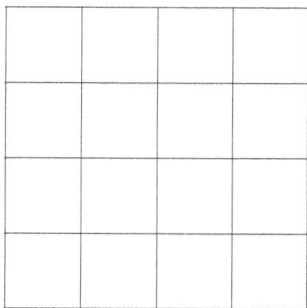

Finder

Date _____   Time _____

Location _____

Telescope _____

Sky Conditions _____

Object _____

Field Drawing

Low Power

Eyepiece:	Mag:
Filter:	FOV:

High Power

Eyepiece:	Mag:
Filter:	FOV:

Notes

_____

_____

_____

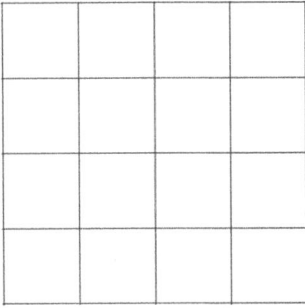

Finder

Date _____   Time _____

Location _____

Telescope _____

Sky Conditions _____

Object _____

Field Drawing

Low Power

High Power

Eyepiece:	Mag:
Filter:	FOV:

Eyepiece:	Mag:
Filter:	FOV:

Notes

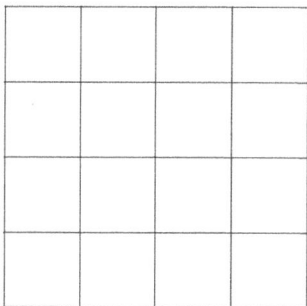

Finder

Date _____ Time _____

Location _____

Telescope _____

Sky Conditions _____

Object _____

Field Drawing

Low Power

High Power

Eyepiece:	Mag:
Filter:	FOV:

Eyepiece:	Mag:
Filter:	FOV:

Notes

_____

_____

_____

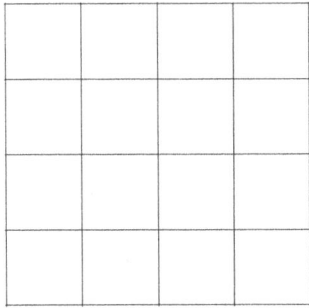

Finder

Date _____   Time _____

Location _____

Telescope _____

Sky Conditions _____

Object _____

Field Drawing

Low Power

High Power

Eyepiece:	Mag:
Filter:	FOV:

Eyepiece:	Mag:
Filter:	FOV:

Notes

_____

_____

_____

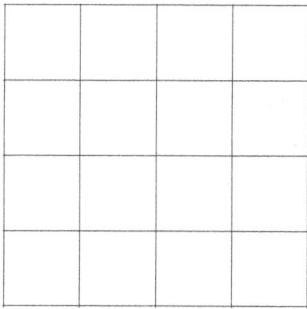

Finder

Date _____ Time _____

Location _____

Telescope _____

Sky Conditions _____

Object _____

Field Drawing

Low Power

High Power

Eyepiece:	Mag:
Filter:	FOV:

Eyepiece:	Mag:
Filter:	FOV:

Notes

_____

_____

_____

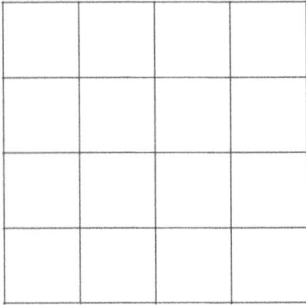

Finder

Date _____   Time _____

Location _____

Telescope _____

Sky Conditions _____

Object _____

Field Drawing

Low Power   High Power

Eyepiece:	Mag:
Filter:	FOV:

Eyepiece:	Mag:
Filter:	FOV:

Notes

_____

_____

_____

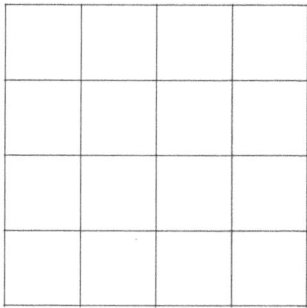

Finder

Date _____   Time _____

Location _____

Telescope _____

Sky Conditions _____

Object _____

Field Drawing

Low Power

High Power

Eyepiece:	Mag:
Filter:	FOV:

Eyepiece:	Mag:
Filter:	FOV:

Notes

_____

_____

_____

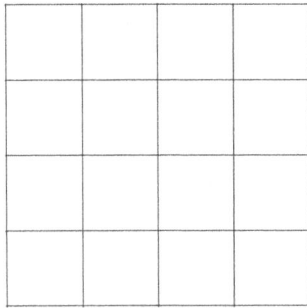

Finder

Date _____   Time _____

Location _____

Telescope _____

Sky Conditions _____

Object _____

Field Drawing

Low Power

High Power

Eyepiece:	Mag:
Filter:	FOV:

Eyepiece:	Mag:
Filter:	FOV:

Notes

_____

_____

_____

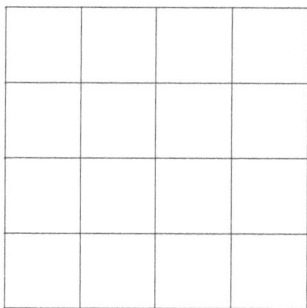

Finder

Date _____   Time _____

Location _____

Telescope _____

Sky Conditions _____

Object _____

Field Drawing

Low Power

Eyepiece:	Mag:
Filter:	FOV:

High Power

Eyepiece:	Mag:
Filter:	FOV:

Notes

_____

_____

_____

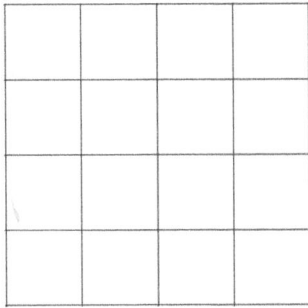

Finder

Date _____ Time _____

Location _____

Telescope _____

Sky Conditions _____

Object _____

Field Drawing

Low Power

High Power

Eyepiece:	Mag:
Filter:	FOV:

Eyepiece:	Mag:
Filter:	FOV:

Notes

_____

_____

_____

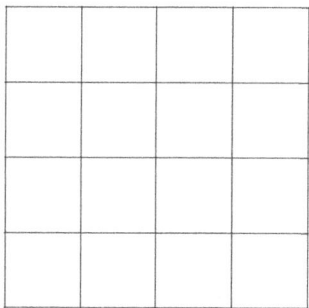

Finder

Date _____   Time _____

Location _____

Telescope _____

Sky Conditions _____

Object _____

Field Drawing

Low Power

High Power

Eyepiece:	Mag:
Filter:	FOV:

Eyepiece:	Mag:
Filter:	FOV:

Notes

_____

_____

_____

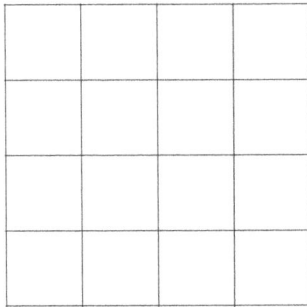

Finder

Date _____ Time _____

Location _____

Telescope _____

Sky Conditions _____

Object _____

Field Drawing

Low Power

High Power

Eyepiece:	Mag:
Filter:	FOV:

Eyepiece:	Mag:
Filter:	FOV:

Notes

_____

_____

_____

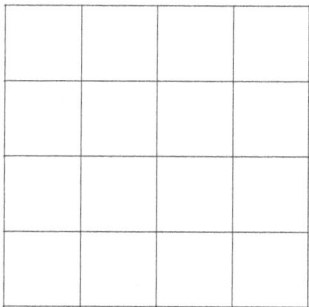

Finder

Date _____   Time _____

Location _____

Telescope _____

Sky Conditions _____

Object _____

Field Drawing

Low Power

High Power

Eyepiece:	Mag:
Filter:	FOV:

Eyepiece:	Mag:
Filter:	FOV:

Notes

_____

_____

_____

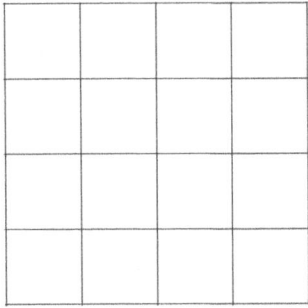

Finder

Date _____   Time _____

Location _____

Telescope _____

Sky Conditions _____

Object _____

Field Drawing

Low Power

High Power

Eyepiece:	Mag:
Filter:	FOV:

Eyepiece:	Mag:
Filter:	FOV:

Notes

_____

_____

_____

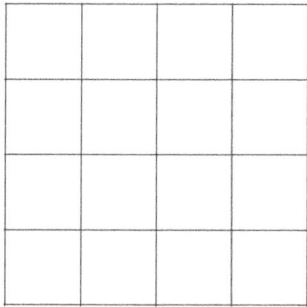

Finder

Date _____ Time _____

Location _____

Telescope _____

Sky Conditions _____

Object _____

Field Drawing

Low Power

Eyepiece:	Mag:
Filter:	FOV:

High Power

Eyepiece:	Mag:
Filter:	FOV:

Notes

_____

_____

_____

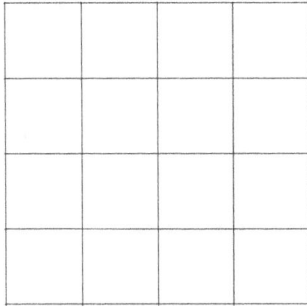

Finder

Date _____   Time _____

Location _____

Telescope _____

Sky Conditions _____

Object _____

Field Drawing

Low Power

High Power

Eyepiece:	Mag:
Filter:	FOV:

Eyepiece:	Mag:
Filter:	FOV:

Notes

_____

_____

_____

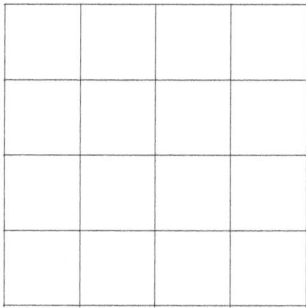

Finder

Date _____   Time _____

Location _____

Telescope _____

Sky Conditions _____

Object _____

Field Drawing

Low Power

High Power

Eyepiece:	Mag:
Filter:	FOV:

Eyepiece:	Mag:
Filter:	FOV:

Notes

_____

_____

_____

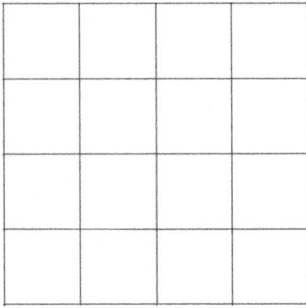

Finder

Date _____   Time _____

Location _____

Telescope _____

Sky Conditions _____

Object _____

Field Drawing

Low Power

High Power

Eyepiece:	Mag:
Filter:	FOV:

Eyepiece:	Mag:
Filter:	FOV:

Notes

_____

_____

_____

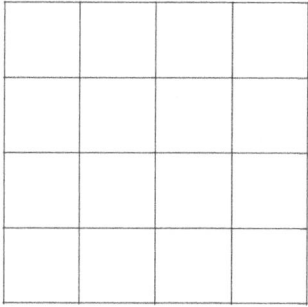

Finder

Date _____ Time _____

Location _____

Telescope _____

Sky Conditions _____

Object _____

Field Drawing

Low Power

Eyepiece:	Mag:
Filter:	FOV:

High Power

Eyepiece:	Mag:
Filter:	FOV:

Notes

_____

_____

_____

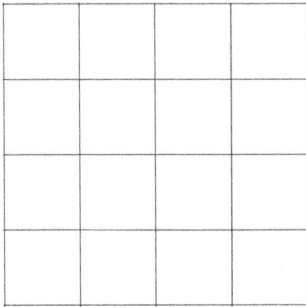

Finder

Date _____   Time _____

Location _____

Telescope _____

Sky Conditions _____

Object _____

Field Drawing

Low Power

High Power

Eyepiece:	Mag:
Filter:	FOV:

Eyepiece:	Mag:
Filter:	FOV:

Notes

_____

_____

_____

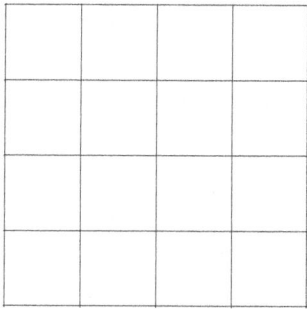

Finder

Date _____   Time _____

Location _____

Telescope _____

Sky Conditions _____

Object _____

Field Drawing

Low Power

Eyepiece:	Mag:
Filter:	FOV:

High Power

Eyepiece:	Mag:
Filter:	FOV:

Notes

_____

_____

_____

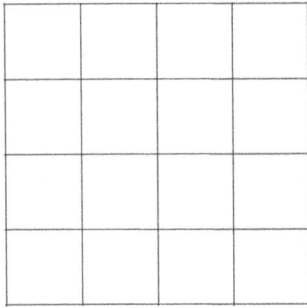

Date _____   Time _____

Location _____

Telescope _____

Sky Conditions _____

Object _____

Finder

Field Drawing

Low Power

High Power

Eyepiece:	Mag:
Filter:	FOV:

Eyepiece:	Mag:
Filter:	FOV:

Notes

_____

_____

_____

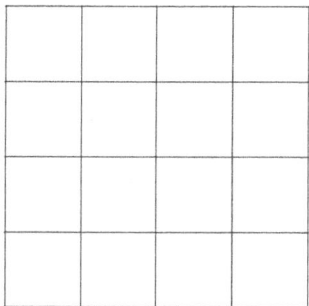

Finder

Date _____   Time _____

Location _____

Telescope _____

Sky Conditions _____

Object _____

Field Drawing

Low Power

High Power

Eyepiece:	Mag:
Filter:	FOV:

Eyepiece:	Mag:
Filter:	FOV:

Notes

_____

_____

_____

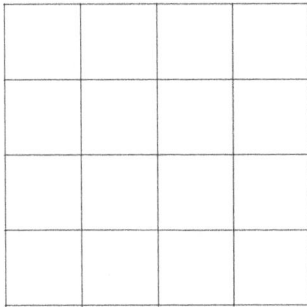

Finder

Date _____   Time _____

Location _____

Telescope _____

Sky Conditions _____

Object _____

Field Drawing

Low Power

High Power

Eyepiece:	Mag:
Filter:	FOV:

Eyepiece:	Mag:
Filter:	FOV:

Notes

_____

_____

_____

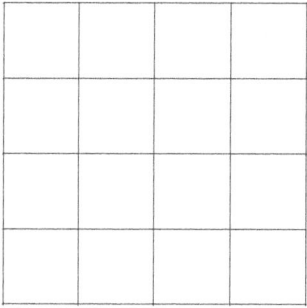

Finder

Date _____ Time _____

Location _____

Telescope _____

Sky Conditions _____

Object _____

Field Drawing

Low Power

High Power

Eyepiece:	Mag:
Filter:	FOV:

Eyepiece:	Mag:
Filter:	FOV:

Notes

_____

_____

_____

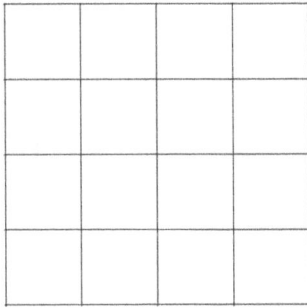

Finder

Date _____ Time _____

Location _____

Telescope _____

Sky Conditions _____

Object _____

Field Drawing

Low Power

High Power

Eyepiece:	Mag:
Filter:	FOV:

Eyepiece:	Mag:
Filter:	FOV:

Notes

_____

_____

_____

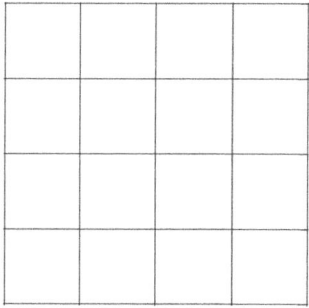

Finder

Date _____   Time _____

Location _____

Telescope _____

Sky Conditions _____

Object _____

Field Drawing

Low Power

High Power

Eyepiece:	Mag:
Filter:	FOV:

Eyepiece:	Mag:
Filter:	FOV:

Notes

_____

_____

_____

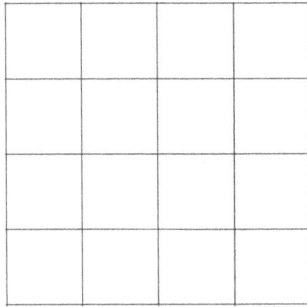

Finder

Date _____   Time _____

Location _____

Telescope _____

Sky Conditions _____

Object _____

Field Drawing

Low Power

High Power

Eyepiece:	Mag:
Filter:	FOV:

Eyepiece:	Mag:
Filter:	FOV:

Notes

_____

_____

_____

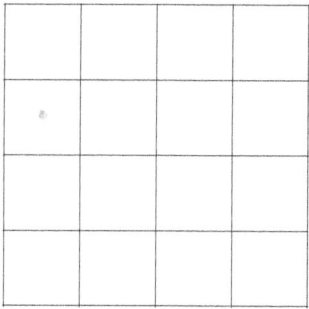

Finder

Date _____   Time _____

Location _____

Telescope _____

Sky Conditions _____

Object _____

Field Drawing

Low Power

High Power

Eyepiece:	Mag:
Filter:	FOV:

Eyepiece:	Mag:
Filter:	FOV:

Notes

_____

_____

_____

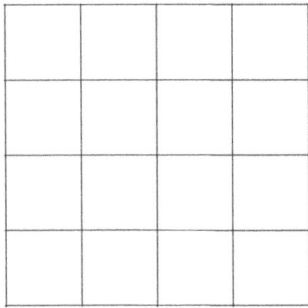

Finder

Date _____   Time _____

Location _____

Telescope _____

Sky Conditions _____

Object _____

Field Drawing

Low Power

High Power

Eyepiece:	Mag:
Filter:	FOV:

Eyepiece:	Mag:
Filter:	FOV:

Notes

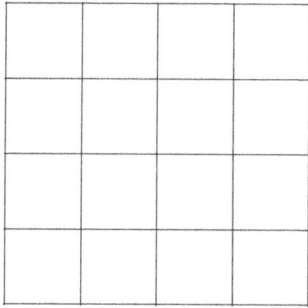

Finder

Date _____   Time _____

Location _____

Telescope _____

Sky Conditions _____

Object _____

Field Drawing

Low Power

High Power

Eyepiece:	Mag:
Filter:	FOV:

Eyepiece:	Mag:
Filter:	FOV:

Notes

_____

_____

_____

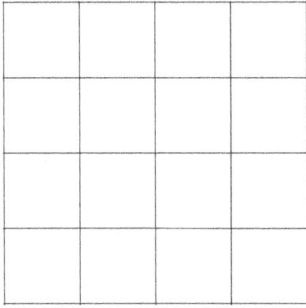

Finder

Date _____   Time _____

Location _____

Telescope _____

Sky Conditions _____

Object _____

Field Drawing

Low Power

Eyepiece:	Mag:
Filter:	FOV:

High Power

Eyepiece:	Mag:
Filter:	FOV:

Notes

_____

_____

_____

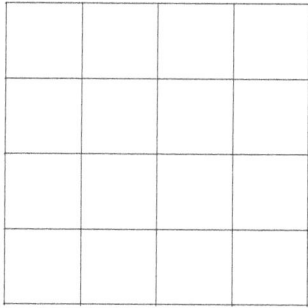

Finder

Date _____   Time _____

Location _____

Telescope _____

Sky Conditions _____

Object _____

Field Drawing

Low Power

Eyepiece:	Mag:
Filter:	FOV:

High Power

Eyepiece:	Mag:
Filter:	FOV:

Notes

_____

_____

_____

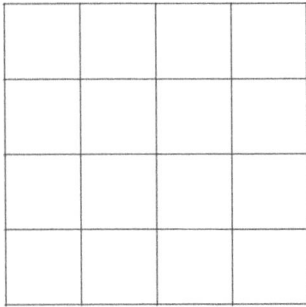

Finder

Date _____   Time _____

Location _____

Telescope _____

Sky Conditions _____

Object _____

Field Drawing

Low Power

High Power

Eyepiece:	Mag:
Filter:	FOV:

Eyepiece:	Mag:
Filter:	FOV:

Notes

_____

_____

_____

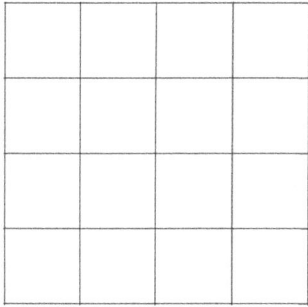

Finder

Date _____ Time _____

Location _____

Telescope _____

Sky Conditions _____

Object _____

Field Drawing

Low Power

High Power

Eyepiece:	Mag:
Filter:	FOV:

Eyepiece:	Mag:
Filter:	FOV:

Notes

_____
_____
_____

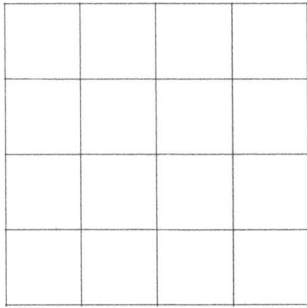

Finder

Date _____   Time _____

Location _____

Telescope _____

Sky Conditions _____

Object _____

Field Drawing

Low Power

High Power

Eyepiece:	Mag:
Filter:	FOV:

Eyepiece:	Mag:
Filter:	FOV:

Notes

_____

_____

_____

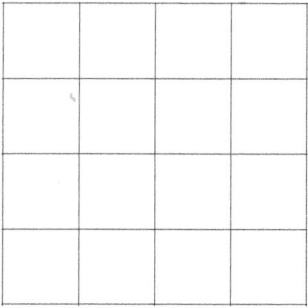

Finder

Date _____   Time _____

Location _____

Telescope _____

Sky Conditions _____

Object _____

Field Drawing

Low Power

High Power

Eyepiece:	Mag:
Filter:	FOV:

Eyepiece:	Mag:
Filter:	FOV:

Notes

_____

_____

_____

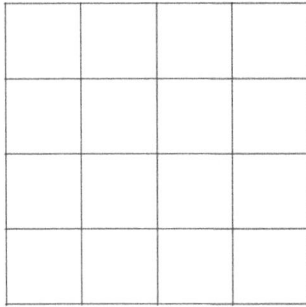

Finder

Date _____   Time _____

Location _____

Telescope _____

Sky Conditions _____

Object _____

Field Drawing

Low Power

High Power

Eyepiece:	Mag:
Filter:	FOV:

Eyepiece:	Mag:
Filter:	FOV:

Notes

_____

_____

_____

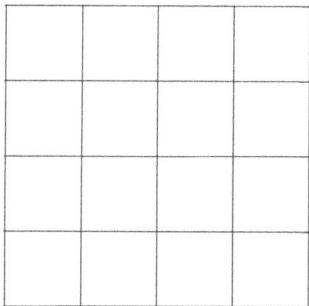

Finder

Date _____ Time _____

Location _____

Telescope _____

Sky Conditions _____

Object _____

Field Drawing

Low Power

High Power

Eyepiece:	Mag:
Filter:	FOV:

Eyepiece:	Mag:
Filter:	FOV:

Notes

_____

_____

_____

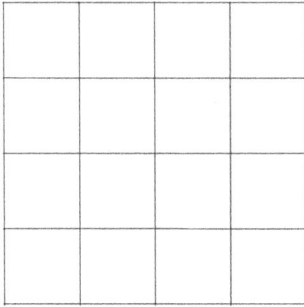

Finder

Date _____   Time _____

Location _____

Telescope _____

Sky Conditions _____

Object _____

Field Drawing

Low Power

Eyepiece:	Mag:
Filter:	FOV:

High Power

Eyepiece:	Mag:
Filter:	FOV:

Notes

_____

_____

_____

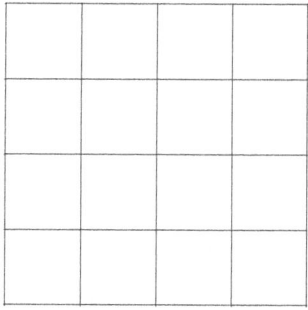

Finder

Date _____ Time _____

Location _____

Telescope _____

Sky Conditions _____

Object _____

Field Drawing

Low Power

High Power

Eyepiece:	Mag:
Filter:	FOV:

Eyepiece:	Mag:
Filter:	FOV:

Notes

_____

_____

_____

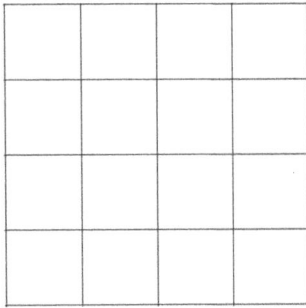

Finder

Date _____   Time _____

Location _____

Telescope _____

Sky Conditions _____

Object _____

Field Drawing

Low Power

High Power

Eyepiece:	Mag:
Filter:	FOV:

Eyepiece:	Mag:
Filter:	FOV:

Notes

_____

_____

_____

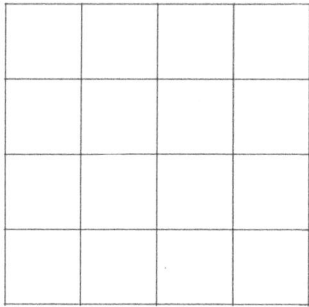

Finder

Date _____   Time _____

Location _____

Telescope _____

Sky Conditions _____

Object _____

Field Drawing

Low Power

Eyepiece:	Mag:
Filter:	FOV:

High Power

Eyepiece:	Mag:
Filter:	FOV:

Notes

_____

_____

_____

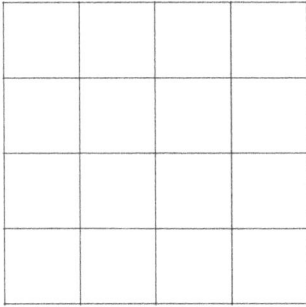

Finder

Date _____   Time _____

Location _____

Telescope _____

Sky Conditions _____

Object _____

Field Drawing

Low Power

High Power

Eyepiece:	Mag:
Filter:	FOV:

Eyepiece:	Mag:
Filter:	FOV:

Notes

_____

_____

_____

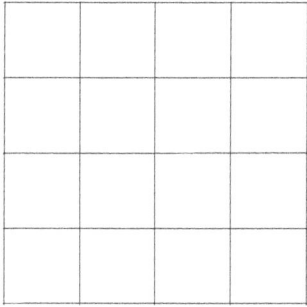

Finder

Date _____   Time _____

Location _____

Telescope _____

Sky Conditions _____

Object _____

Field Drawing

Low Power

Eyepiece:	Mag:
Filter:	FOV:

High Power

Eyepiece:	Mag:
Filter:	FOV:

Notes

_____

_____

_____

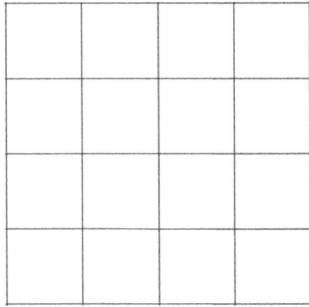

Finder

Date _____   Time _____

Location _____

Telescope _____

Sky Conditions _____

Object _____

Field Drawing

Low Power

High Power

Eyepiece:	Mag:
Filter:	FOV:

Eyepiece:	Mag:
Filter:	FOV:

Notes

_____

_____

_____

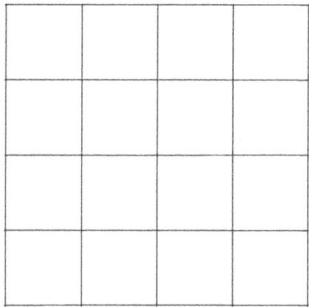

Finder

Date _____ Time _____

Location _____

Telescope _____

Sky Conditions _____

Object _____

Field Drawing

Low Power

High Power

Eyepiece:	Mag:
Filter:	FOV:

Eyepiece:	Mag:
Filter:	FOV:

Notes

_____

_____

_____

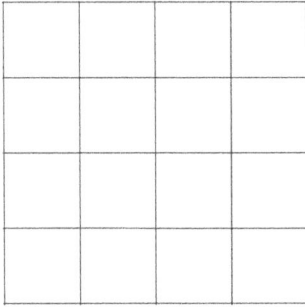

Finder

Date _____   Time _____

Location _____

Telescope _____

Sky Conditions _____

Object _____

Field Drawing

Low Power

Eyepiece:	Mag:
Filter:	FOV:

High Power

Eyepiece:	Mag:
Filter:	FOV:

Notes

_____

_____

_____

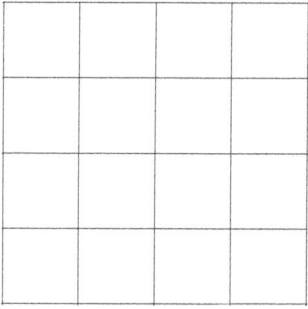

Finder

Date _____   Time _____

Location _____

Telescope _____

Sky Conditions _____

Object _____

Field Drawing

Low Power

Eyepiece:	Mag:
Filter:	FOV:

High Power

Eyepiece:	Mag:
Filter:	FOV:

Notes

_____

_____

_____

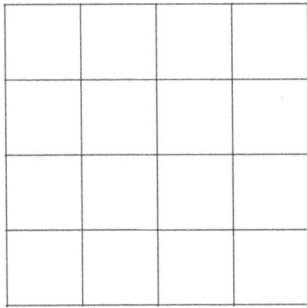

Finder

Date _____   Time _____

Location _____

Telescope _____

Sky Conditions _____

Object _____

Field Drawing

Low Power

High Power

Eyepiece:	Mag:
Filter:	FOV:

Eyepiece:	Mag:
Filter:	FOV:

Notes

_____

_____

_____

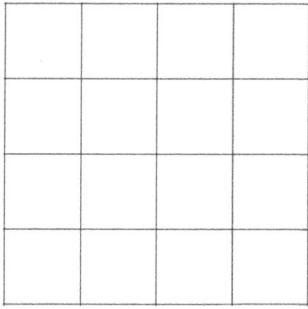

Finder

Date _____   Time _____

Location _____

Telescope _____

Sky Conditions _____

Object _____

Field Drawing

Low Power

High Power

Eyepiece:	Mag:
Filter:	FOV:

Eyepiece:	Mag:
Filter:	FOV:

Notes

_____

_____

_____

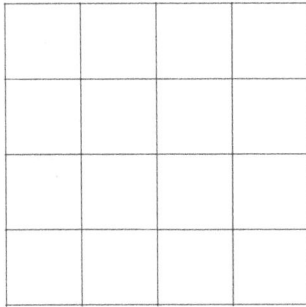

Finder

Date _____   Time _____

Location _____

Telescope _____

Sky Conditions _____

Object _____

Field Drawing

Low Power

High Power

Eyepiece:	Mag:
Filter:	FOV:

Eyepiece:	Mag:
Filter:	FOV:

Notes

_____

_____

_____

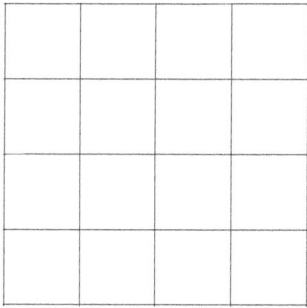

Finder

Date _____   Time _____

Location _____

Telescope _____

Sky Conditions _____

Object _____

Field Drawing

Low Power

High Power

Eyepiece:	Mag:
Filter:	FOV:

Eyepiece:	Mag:
Filter:	FOV:

Notes

_____

_____

_____

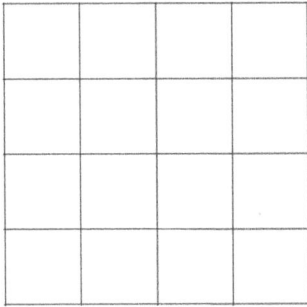

Finder

Date _____ Time _____

Location _____

Telescope _____

Sky Conditions _____

Object _____

Field Drawing

Low Power

High Power

Eyepiece:	Mag:
Filter:	FOV:

Eyepiece:	Mag:
Filter:	FOV:

Notes

_____

_____

_____

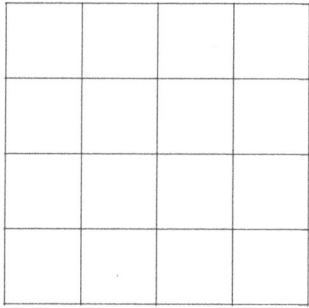

Finder

Date _____   Time _____

Location _____

Telescope _____

Sky Conditions _____

Object _____

Field Drawing

Low Power

Eyepiece:	Mag:
Filter:	FOV:

High Power

Eyepiece:	Mag:
Filter:	FOV:

Notes

_____

_____

_____

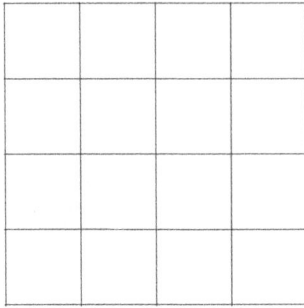

Finder

Date _____   Time _____

Location _____

Telescope _____

Sky Conditions _____

Object _____

Field Drawing

Low Power

Eyepiece:	Mag:
Filter:	FOV:

High Power

Eyepiece:	Mag:
Filter:	FOV:

Notes

_____

_____

_____

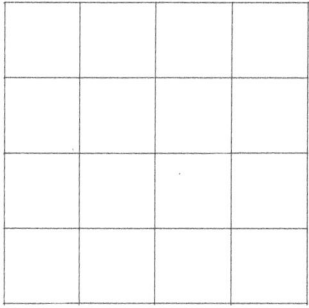

Finder

Date _____   Time _____

Location _____

Telescope _____

Sky Conditions _____

Object _____

Field Drawing

Low Power

Eyepiece:	Mag:
Filter:	FOV:

High Power

Eyepiece:	Mag:
Filter:	FOV:

Notes

_____

_____

_____

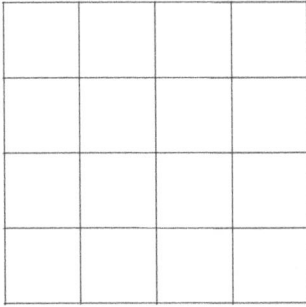

Finder

Date _____ Time _____

Location _____

Telescope _____

Sky Conditions _____

Object _____

Field Drawing

Low Power

High Power

Eyepiece:	Mag:
Filter:	FOV:

Eyepiece:	Mag:
Filter:	FOV:

Notes

_____

_____

_____

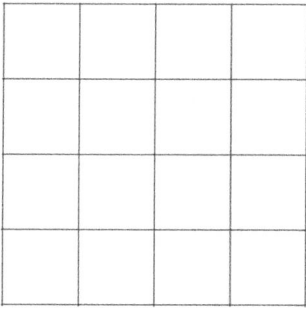

Finder

Date _____   Time _____

Location _____

Telescope _____

Sky Conditions _____

Object _____

Field Drawing

Low Power

High Power

Eyepiece:	Mag:
Filter:	FOV:

Eyepiece:	Mag:
Filter:	FOV:

Notes

_____

_____

_____

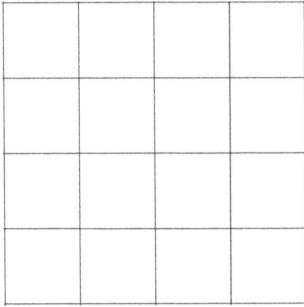

Finder

Date _____   Time _____

Location _____

Telescope _____

Sky Conditions _____

Object _____

Field Drawing

Low Power

High Power

Eyepiece:	Mag:
Filter:	FOV:

Eyepiece:	Mag:
Filter:	FOV:

Notes

_____

_____

_____

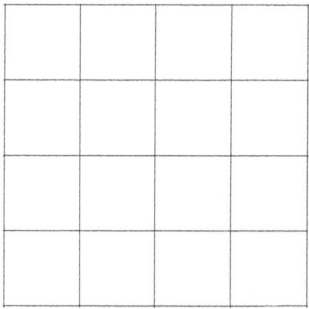

Finder

Date _____   Time _____

Location _____

Telescope _____

Sky Conditions _____

Object _____

Field Drawing

Low Power

Eyepiece:	Mag:
Filter:	FOV:

High Power

Eyepiece:	Mag:
Filter:	FOV:

Notes

_____

_____

_____

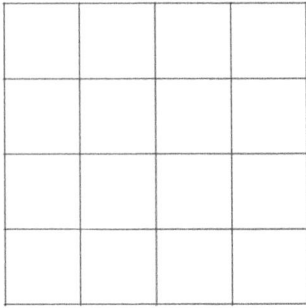

Finder

Date _____   Time _____

Location _____

Telescope _____

Sky Conditions _____

Object _____

Field Drawing

Low Power

High Power

Eyepiece:	Mag:
Filter:	FOV:

Eyepiece:	Mag:
Filter:	FOV:

Notes

_____

_____

_____

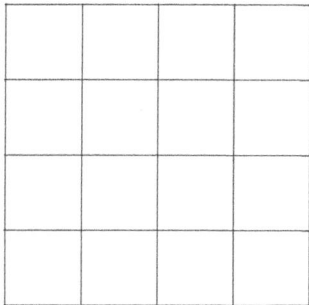

Finder

Date _____   Time _____

Location _____

Telescope _____

Sky Conditions _____

Object _____

Field Drawing

Low Power

High Power

Eyepiece:	Mag:
Filter:	FOV:

Eyepiece:	Mag:
Filter:	FOV:

Notes

_____

_____

_____

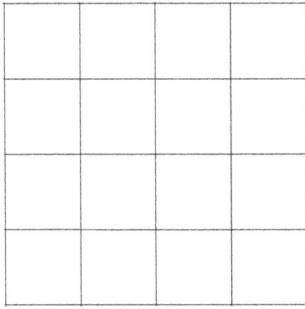

Finder

Date _____ Time _____

Location _____

Telescope _____

Sky Conditions _____

Object _____

---------- **Field Drawing** ----------

Low Power   High Power

Eyepiece:	Mag:
Filter:	FOV:

Eyepiece:	Mag:
Filter:	FOV:

---------- **Notes** ----------

_____

_____

_____

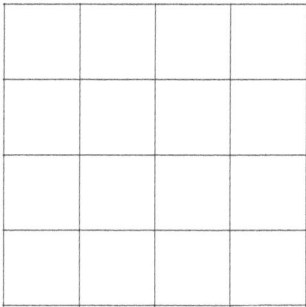

Finder

Date _____ Time _____

Location _____

Telescope _____

Sky Conditions _____

Object _____

Field Drawing

Low Power

High Power

Eyepiece:	Mag:
Filter:	FOV:

Eyepiece:	Mag:
Filter:	FOV:

Notes

_____

_____

_____

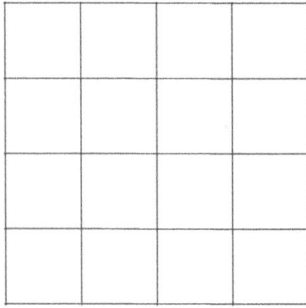

Finder

Date _____   Time _____

Location _____

Telescope _____

Sky Conditions _____

Object _____

Field Drawing

Low Power

High Power

Eyepiece:	Mag:
Filter:	FOV:

Eyepiece:	Mag:
Filter:	FOV:

Notes

_____

_____

_____

www.ingramcontent.com/pod-product-compliance
Lightning Source LLC
Chambersburg PA
CBHW080559030426
42336CB00019B/3258